Great Ideas of Science

RELATIVITY

by Judith Herbst

Twenty-First Century Books
Minneapolis

Twenty-First Century Books
A division of Lerner Publishing Group
241 First Avenue North
Minneapolis, Minnesota 55401 U.S.A.

Website address: www.lernerbooks.com

Library of Congress Cataloging-in-Publication Data

Herbst, Judith.
    Relativity / Judith Herbst.
        p.   cm. — (Great ideas of science)
    Includes bibliographical references and index.
    ISBN-13: 978–0–8225–2918–7 (lib. bdg. : alk. paper)
    ISBN-10: 0–8225–2918–1 (lib. bdg. : alk. paper)
    1. Relativity (Physics) 2. General relativity (Physics) 3. Special
    relativity (Physics) I. Title. II. Series.
    QC173.55.H47 2007
    530.11—dc22   2005008805

Manufactured in the United States of America
1 2 3 4 5 6 – DP – 12 11 10 09 08 07

TABLE OF CONTENTS

INTRODUCTION

You are floating weightlessly in space. The ship from which you were ejected is long gone, swallowed up by the immense distance that now separates you. You have been out here for about an hour, but with no chance of rescue, you wonder if it even matters how much time has passed.

The universe is breathtaking, ablaze with stars, but still so very dark and empty. You seem to be in the center of it all, but you know that's just an illusion. Someone 1,000 light-years away from you would think she was in the center. The people in the spaceship who set you adrift probably think they're in the center. But there is no center. And as you turn and tumble, there is no up or down anymore, no left or right.

All at once, you find yourself in the midst of a wispy gas cloud, but you have no sensation of movement. Is the cloud streaming toward you, or are you passing through it? It's impossible to tell. Far away from the familiar signposts and measuring sticks, the universe at last reveals its true self. It is a universe the ancients could barely imagine.

Until the twentieth century, people believed that everything circled Earth. The stars, Sun, and planets rose in the east and set in the west, while Earth remained at rest in the middle of it all. Time was a river that flowed only one way, and space had three dimensions. Apples fell from trees because a force called gravity pulled them down. It was obvious which objects were in motion and which were not.

But then along came the great physicist Albert Einstein, who rocked the scientific world with his special and general theories of relativity. Einstein published his

Albert Einstein drastically changed our understanding of physics when he published his special and general theories of relativity in the early 1900s. He is pictured here around the time that he published his relativity theories.

special theory of relativity in 1905. In it he explained that time and motion are relative to the observer, rather than fixed and absolute. Space and time are not separate entities, but linked in the reality called space-time. Mass and energy are two forms of the same thing, and mass can be converted into huge amounts of energy. In 1916 Einstein published his theory of general relativity, in which he showed that gravity is not a force, but a warp in the fabric of space that is created by the presence of mass.

Few theories have made our eyes widen in astonishment like relativity. Its concepts are not at all what we expect and seem so contrary to what we see everyday. But that is what makes relativity such an important theory. Einstein made a leap, not just of thought, but of experience. He stepped out of the known into the vastness of space.

CHAPTER 1

THE ROAD TO RELATIVITY

Common sense told the ancients that Earth doesn't move. If it did, they would surely have felt it. But it was quite clear that just about everything else in the sky moved. Over time, the Moon, Sun, and planets all changed their positions, while Earth remained motionless in the middle. This seemed so obvious that for a long time nobody even thought to question it. One of the first to do so was a Greek named Philolaus.

Philolaus was born in Italy sometime around 480 B.C. He wasn't a scientist, but a philosopher, so his suggestion that Earth was not stationary and not in the center of everything may have simply been a lucky guess. But lucky guess or not, Philolaus's picture of the universe horrified just about everyone. The ancient Greeks saw the heavens as perfect and unchanging. Earth was in a regal central position, like a king upon a throne, with all the lesser bodies humbly circling it. So Philolaus's radical views were ignored.

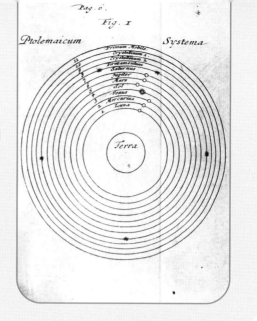

An early diagram of the universe *(right)* depicts Earth (labeled Terra) at the center, with the Moon, Sun, and other planets in orbit around it.

About one hundred years later, the Greek astronomer Heracleides made a very interesting observation. It had always been assumed that the vault of heaven—the dark crystal sphere in which the stars were thought to be embedded—turned as a single unit while Earth stood still. But Heracleides realized that we would see exactly the same movement if the heavens stood still and Earth turned. This was only partially right, as he left Earth in the center, but he did recognize that motion might be relative—it might depend on the observer's perspective.

Meanwhile, Aristotle, a Greek philosopher, said that Heracleides' theory was nonsense. He argued that if Earth were turning, you wouldn't be able to throw a ball into the air and then catch it when it fell back down. The spinning Earth would carry you away from it. At the time, this seemed perfectly logical. But there was a flaw in Aristotle's reasoning.

If Earth is moving, then we, who are standing on Earth, must be moving with it. The air that surrounds us and the ball that has been tossed into it are also being carried along. Because everything is moving together, it seems as if nothing is moving. This is one of the reasons why the concept of a moving Earth was so hard to accept. You couldn't feel it, and you couldn't prove it.

The Greek astronomer Aristarchus lived in the third century B.C. He theorized that the Sun is in the center of the solar system, with Earth and the other planets in motion around it. In addition, Aristarchus said Earth makes one complete turn on its axis every twenty-four hours as it circles the Sun.

The Polish astronomer Copernicus came along in the middle of the fifteenth century A.D. At that time, the Church of Rome had great political power. The church's official belief was that Earth was the center of the universe. Anyone who said Earth moved was questioning church doctrine, and that was risky. Copernicus knew this, but he nevertheless published the details of his Sun-centered model of the solar system. His theory correctly placed Earth's orbit beyond those of Mercury and Venus. A few astronomers quickly embraced the new solar system model, although it would not gain general acceptance until long after Copernicus's death.

One hundred years later, the great Italian astronomer Galileo Galilei was forced by the church to renounce his own Sun-centered theory and pay penance by reciting psalms once a week for three years. But the punishment didn't change Galileo's mind. According to legend, each

When Galileo Galilei *(left, with cape)* published his *Dialogue Concerning the Two Chief World Systems* about Copernicus's Sun-centered theory in 1632, the pope placed Galileo under house arrest.

time Galileo rose from his recitation, he murmured, "Eppur si muove!" (And yet it moves.)

NEWTON'S LAWS OF MOTION

Isaac Newton, one of the greatest scientists who ever lived, was born in England in 1642. Newton puzzled over many things: the motion of objects; the structure of space; and the nature of light, time, and gravity. When he was in his twenties, he developed the principles of calculus. These mathematical methods can be used to solve a variety of physics problems. Newton is probably best known for his three laws of motion. These laws describe how forces affect the motion of objects.

Newton's first law says that a stationary object won't move unless an outside force sets it in motion. This reluctance to move is called inertia, from a Latin word meaning "unskilled" or "inactive." But once that object is in motion, it will keep going in a straight line unless an outside force stops it or makes it change direction.

Let's suppose your dog nudges his favorite rubber ball across the floor with his nose. If the ball doesn't hit the leg of the coffee table or crash into the wall, it will roll in a straight line. But friction between the ball and the floor will slow it down, and eventually it will stop rolling.

What if Quark, the cosmic cat, bats a sponge ball? The vacuum of space has no friction. Therefore the ball will travel in a straight line at a steady speed forever, as long as nothing gets in its way. But is the ball *really* moving in a straight line?

Suppose we thread a very long, stiff wire through Quark's sponge ball. Quark bats the ball again and sets it in motion along the wire. Since the wire is straight and cannot bend, we know that the ball is moving in a straight line. But Quark is a very playful cosmic cat. He picks up the wire with his mouth and begins to zoom through space with it, shaking his head this way and that. The ball is still moving in a straight line along the straight wire, but the wire is not moving in a straight line. Thanks to Quark's antics, it is zigzagging right and left, and up and down. So is the ball moving in a straight line, or isn't it?

And there's another problem. Quark's ball didn't just materialize next to his paw. It already existed as an object in space, like a planet or a bit of cosmic dust. Quark came across it during his romp through the universe and set it in motion. But was the ball ever at rest in the first place?

The idea of something being at rest was a problem for Newton. He knew that everything in space moved. Earth and the planets orbited the Sun, moons orbited planets, and stars moved through the universe. Nothing was standing still. So how could he speak about objects that were at rest? The idea of an object moving in a straight line was also a bit murky. A straight line according to what?

NEWTON'S THREE LAWS OF MOTION

Newton's first law says that an object at rest will remain at rest unless it is set in motion. An object that is in motion will remain in motion in a straight line, neither speeding up or slowing down, until something acts on it and causes it to change speed or direction.

In his second law, Newton states that the change of motion is proportional to the applied force and occurs in the same direction as that force. Imagine pushing a wheelbarrow. Newton's law says that the wheelbarrow will move in the direction in which you push it. If you push harder, it will move faster.

Newton's third law is often called the equal and opposite law. It states that for every action there is an equal and opposite reaction. So suppose you are in a rowboat on a lake and decide to take a swim. You stand on the bench at the rear of the boat and jump into the water. You go one way, and the force of your dive causes the boat to move in the opposite direction.

Newton solved the problem of relative motion by saying that space itself is fixed and unchanging. In a universe of motion, space remains the great ruler against which all motion can be measured. You can imagine Newton's space as the area inside an inflated balloon that never loses air. The balloon's volume and shape are constant. So objects can be at rest *in* space or move in a straight line *through* space.

NEWTON AND GRAVITY

When the Great Plague spread across Europe in 1665, Newton had just finished his studies at Cambridge University in London. Everyone who could fled the city to escape this disease that no

Sir Isaac Newton

one could cure. Newton went to his mother's farm in the quiet village of Woolsthorpe. According to legend, one day Newton settled down under a tree to think. In the midst of his pondering, an apple fell to the ground. Newton was struck, not by the apple, but by an amazing realization. The force that pulled the apple toward the ground was the same force that held the Moon in orbit around Earth!

Ever since Aristotle, scientists and philosophers had believed in two different sets of natural laws. One set governed what happened on Earth, and the other de-scribed events in the heavens. But Newton realized that everything in the universe operates according to a single set of physical laws.

Galileo had shown that all falling objects accelerate at the same rate of speed, regardless of their weight. Newton understood that this acceleration was due to gravity. He realized that gravity was a force of attraction between two objects and that the larger object pulled the smaller one toward it. He began to wonder how far grav-ity's effects extended. Could gravity's influence reach be-yond Earth, to the Moon and beyond? He calculated the force needed to keep the Moon in its orbit around Earth and compared it with the force that pulled the apple downward. After taking into account the Moon's greater distance from Earth and its larger mass, he discovered that the forces were the same. The Moon was held in its orbit by Earth's gravity.

Newton's calculations showed that the force of gravity decreases as the inverse square of the distance. If you double the distance, the force of gravity is reduced to one-fourth of what it was originally. If the distance is tripled, gravity is reduced to one-ninth of what it was originally. The farther away two objects are, the less they are affected by each other's gravitational pull.

Newton knew that gravity acts instantaneously. He knew that it operates in a vacuum, needing nothing to "carry" it, unlike sound waves that need a medium such as air to carry them. He understood how gravity affects the motion of objects, but not why it works the way it does.

WHY DOESN'T THE MOON FALL OUT OF THE SKY?

Earth's gravity holds the Moon in its orbit. But if the Moon is close enough to be affected by Earth's gravity, why doesn't it fall down, the way ripe apples fall from trees?

The secret of the Moon's balancing act has to do with its forward motion. Newton's first law says that the Moon should travel in a straight line unless an outside force acts on it. So picture the Moon traveling through space on a straight path that is perpendicular to a line between it and Earth's center. At the same time, Earth's gravity is trying to pull the Moon straight down. The resulting movement is a compromise. The Moon neither travels in a straight line nor falls straight down. It falls in a curved path around Earth. The same holds true for planets orbiting stars.

ELECTRICITY AND MAGNETISM

By the beginning of the nineteenth century, scientists suspected that electricity and magnetism were related in some way. But no one had yet been able to prove it. Then, in 1819, the Danish physicist Hans Christian Oersted placed a compass next to a wire through which an electric current was flowing. The compass's needle moved until it pointed at right angles to the wire. When he reversed the current, the needle turned in the opposite direction. The electric current flowing through the wire had created a magnetic field that deflected the needle. Oersted had proven a connection between electricity and magnetism.

Oersted published the results of his experiment the

following year. His findings ignited the scientific community. They especially piqued the interest of a young Englishman named Michael Faraday.

Faraday was working for Humphry Davy, a famous chemist, when he read Oersted's paper showing that an electric current could create a magnetic field. Faraday wondered if the reverse might be true. So he set up his own experiment. He wound a piece of wire around each side of an iron ring. He connected one coil of wire to a galvanometer—a device that measures small electric currents—and the other to a battery. Faraday expected the electric current to magnetize the iron ring. He hoped that the magnetized ring would induce an electric current in the second coil of wire. As a current began to flow from the battery, the galvanometer briefly detected a current in the second wire. When Faraday disconnected the battery, the galvanometer once again detected a burst of current in the second wire. The magnetized ring had induced an electric current—but only a temporary one.

Faraday used visualization to explain what happened during his experiment. He imagined that when an electric current is turned on, it creates a magnetic force field that spreads out in all directions. He had, in fact, "seen" such an invisible magnetic field. He had sprinkled iron filings on a piece of paper under which he had placed a magnet. The bits of iron had positioned themselves along what Faraday called lines of force. The lines connected points of equal field strength.

In Faraday's magnetic induction experiment, the current passing through the first wire magnetized the iron ring. Faraday speculated that the induced magnetic field sent out lines of force that crossed the second wire and briefly caused an electric current to flow through it. When Faraday switched off the current in the first wire, the field lines around the iron ring collapsed. They again crossed the second wire, causing another quick burst of current.

It was an extraordinary description: invisible magnetic fields and lines of force that formed and collapsed. But Faraday could not prove that these things really existed outside his imagination.

MAXWELL'S ELECTROMAGNETIC THEORY

The Scottish mathematician and physicist James Clerk Maxwell embraced Faraday's theory about the lines of force. He eventually worked out mathematical equations that described in numbers what Faraday had done with words. He also showed mathematically that electricity and magnetism cannot exist as two separate phenomena. Maxwell's work came to be known as the theory of electromagnetism.

But Maxwell wasn't finished. He showed that an oscillating (vibrating) electric charge produced a magnetic field that radiated outward, somewhat like the ripples in a pond when a stone is tossed into the water. This electromagnetic radiation traveled at the speed of light. Maxwell wondered if light itself might be the result of an oscillating electric charge. In other words, was light a form of electromagnetic radiation?

Maxwell further suggested that since electric charges can oscillate at different frequencies, there was probably a whole range of electromagnetic radiation, of which visible light was just one kind. Scientists had already discovered infrared and ultraviolet light, both of which are invisible to the human eye. What other kinds of electromagnetic radiation might exist? Maxwell had predicted what would come to be known as the electromagnetic spectrum.

WAVELENGTHS, FREQUENCIES, AND THE ELECTROMAGNETIC SPECTRUM

Electromagnetic radiation is energy. When an electromagnetic field oscillates, it produces waves that travel outward from the source, carrying energy. Electromagnetic waves have both a wavelength and a frequency. A wave's length is the distance from one peak (highest point) to the next. If a wave is bunched together, its wavelength is said to be short because there's only a short distance between the peaks. If a wave is more spread out, it is said to have a long wavelength. Frequency is how fast an electromagnetic field oscillates. The faster the oscillation, the higher the frequency and the more energetic the radiation.

At one end of the electromagnetic spectrum are the radio waves. They have the longest wavelengths, the lowest frequencies, and the least energy. At the other end is gamma radiation. Gamma rays have the shortest wavelengths, the highest frequencies, and the most energy.

Electromagnetic radiation has a split personality. It can manifest itself as either a wave or a stream of particles. For example, light can be either waves or tiny packets of energy called photons. All electromagnetic radiation travels through space at the same speed—the speed of light.

LIGHT AND MOTION

Since ancient times, scientists and philosophers had thought about the speed of light. Some, like Aristotle, thought the speed of light was infinite. Others believed it was finite and could be measured.

In 1639 Galileo suggested that light might only *appear* to travel instantaneously because it moves so incredibly fast. He set up an experiment to test this idea. He sent two men with lanterns to the tops of two hills located 1.5 miles (2.4 kilometers) apart. Each man covered his lantern, so neither could see the other's light. Then the first man uncovered his light. As he did so, he noted the time. The second man had been instructed to uncover his lantern at the exact moment he saw the light, thereby signaling the first man. When the first man saw the second man's light, he was to note the time. Galileo figured that dividing the round-trip distance by the elapsed time should give the speed of light. Unfortunately, the light traveled much too fast to measure. The speed of light remained unknown.

A computer-enhanced image of Jupiter, made from three black-and-white photos taken in 1979 by *Voyager 1*, shows three of Jupiter's moons. They are *(from left to right)* Callisto, Io, and Europa.

A FINITE BUT VERY FAST VELOCITY

The seventeenth-century Danish astronomer Olaus Roemer spent many hours tracking the motion of Jupiter's four known moons, Io, Europa, Ganymede, and Callisto. He knew how long it took the moons to orbit Jupiter. He figured it should be possible to predict the moons' eclipses—the exact moment when each moon would pass behind the giant planet and be hidden from his view. But the timing of the eclipses didn't match his calculations. At some times of the year, the eclipses took place earlier than he expected. At other times, they were later than he expected. Roemer noticed that the eclipses were ahead of schedule when Earth was closer to Jupiter and behind schedule when Earth and Jupiter were farther apart.

Roemer knew that the light reflected from the moons had to travel farther to reach his eyes when the two planets were farther apart than when they were closer together. If, he reasoned, the light was arriving at different times depending on the distance between the planets, then light clearly does not travel instantaneously. It has a finite speed.

Roemer then repeated Galileo's experiment. But instead of using nearby hills, he used two locations that were millions of miles apart—Earth and Jupiter. Instead of lanterns, Roemer used the flash of light that occurred at the moment of the moon Io's eclipse. Knowing how much Io's timing seemed to change from month to month and how much the distance from Earth to Jupiter varied, Roemer was able to calculate a value for the speed of light: about 140,000 miles (225,000 km) per second. Considering the equipment available to Roemer, his calculation was amazingly close to the modern-day measurement of 186,282 miles (300,000 km) per second.

For the next two hundred years, scientists continued to measure the speed of light, revising and refining it. In 1882

Roemer's Speed-of-Light Experiment

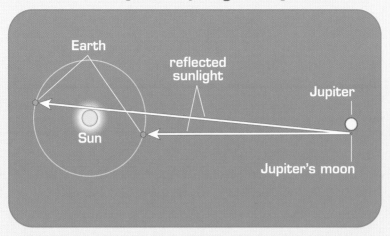

Eclipses of Jupiter's moons seem to come early when Earth and Jupiter are closest together and late when Earth and Jupiter are farthest apart. This is because sunlight reflected from Jupiter's moons must travel farther when Earth and Jupiter are farther apart.

the German American physicist Albert Michelson clocked light traveling through a vacuum at 186,320 miles (300,000 km) per second. This was a remarkable achievement. Michelson was off by a mere 38 miles (61 km) per second.

THE LIGHT-YEAR The fixed speed of light provides us with the astronomical unit of measurement called the light-year. One light-year is the distance light can cover in one year—approximately 5.8 trillion miles (9.3 trillion km). The Sun, at an average distance of 93 million miles (150 million km) from Earth, is just 8 light-*minutes* away. Proxima Centauri, our nearest stellar neighbor (after the Sun), is 4.3 light-years away. Betelgeuse, a bright red star in the constellation Orion, is 652 light-years away.

THE MYSTERIOUS ETHER

We take light for granted. We flip a switch and light appears. But what exactly is light?

The first comprehensive theory of light was presented in 1690 by the Dutch physicist Christiaan Huygens, who said light behaves like waves. In 1704 Isaac Newton proposed that light is composed of tiny particles. A spirited debate followed, as each theory explained certain aspects of light's behavior. Newton's particle theory was favored for many years, but in the 1800s, most scientists came to accept the wave theory.

A problem with the wave theory was that waves need a medium, such as air or water, in which to travel. Consider sound waves. Suppose the evil Dr. Zorg decides to blow up Pluto. He blasts it to smithereens with the latest in death ray technology. On a scale of zero to ten, with ten being deafening, how loud is the explosion? The answer is zero. The explosion occurs in total silence because there is no air in space to carry the sound waves.

Nineteenth-century scientists reasoned that if light is a wave, it has to travel through something. That's when they came up with the idea of the luminiferous ether. The ether was supposedly weightless, transparent, and frictionless. It was everywhere, spreading through all matter and filling all of space with its ability to carry light waves. Many highly regarded scientists of the day believed in the ether. Maxwell, for example, described magnetic lines of force as disturbances in the ether.

In the late 1800s, Albert Michelson *(left)* and Edward Morley *(right)* tested the notion that the universe contained an invisible ether that conducted light waves.

In 1887 American chemist Edward Morley teamed up with Albert Michelson to try to detect Earth's motion through the ether. In their experiment, Michelson and Morley used a device called an interferometer, which can split a beam of light into two beams traveling in different directions. The interferometer was set up so that when the light beam was split, half of it moved parallel to Earth's motion, and the other half moved perpendicular to it. Mirrors reflected both beams to a detector. The scientists compared the travel times of the two halves of the beam. They reasoned that a difference in speed would prove that Earth was indeed moving relative to the ether. As it turned out, both halves of the split light beam trav-

The Michelson-Morley Experiment

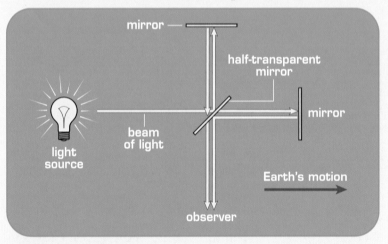

A beam of light travels from the source to the half-transparent mirror. Half of the light is reflected toward the top mirror, and half passes straight through to the other mirror. If light is a wave traveling through the ether, then the speed of the beam that is traveling in the same direction as Earth should be changed by Earth's velocity, but the other one should not.

eled at the same velocity. If the ether existed, Earth was moving with it, not relative to it.

FIXED AND UNCHANGING

Albert Einstein published five scientific papers in 1905. One dealt with the photoelectric effect, a phenomenon in which electrons are ejected from a metal surface when it is exposed to electromagnetic radiation. It was in this paper that Einstein introduced his revolutionary theory of light.

Einstein said that the speed of light is the same for all observers, regardless of what they are doing. Suppose we give Superman a flashlight and ask him to fly through the universe faster than a speeding bullet. Off he goes at a velocity of 20 miles (30 km) per second, with the flashlight's beam shining ahead of him. How fast is the light from the flashlight moving?

It seems logical that the light's speed and Superman's speed should be added together. But this is not the case. Superman can fly as fast as he likes, but his speed will not affect the speed of the light coming from the flashlight. The speed of light traveling through a vacuum will always be 186,282 miles (300,000 km) per second.

Superman, who is carrying the flashlight, measures the beam traveling at 186,282 miles per second. An observer who is standing on Earth, watching Superman shoot by, will also measure the flashlight beam traveling at 186,282 miles per second. The speed of light is not relative. It is, instead, the universal constant.

Five years before Einstein published his paper, the German physicist Max Planck had suggested that energy, like matter, consisted of tiny particles. He called these particles quanta, from a Latin word meaning "how much." Einstein agreed. He concluded that light doesn't need a medium to carry it. But what of light's wavelike properties? The answer, he said, was that light sometimes acts like a stream of particles, and sometimes it acts like a wave.

RELATIVE MOTION

Light moves at a fixed speed. The movement of everything else is relative to the observer. This was the essence of Einstein's first paper on special relativity.

Imagine yourself on a train that is moving west. The train's speed never changes, and the ride is very smooth.

You settle into your seat and turn your head to look out the window. Suddenly you have a strange sensation. The train seems to be at rest, while the scenery outside seems to be rushing past you in the opposite direction. The countryside appears to be moving east, as if it is on a conveyor belt. What is in motion, and what's standing still? This is the classic relativity question. Einstein answered it by saying that when the motion is uniform, with no acceleration nor deceleration, it's impossible to tell what is moving and what is at rest.

Common sense tells you that trains move and the landscape stands still. But Earth itself has four different motions. It turns on its axis, it circles the Sun, it travels around the Milky Way galaxy with the solar system, and it is carried along with the galaxy through the universe. So the landscape isn't standing still. Even the train tracks are in motion! Einstein's special theory of relativity tells us that when we speak of something being "at rest" or "in motion," we must always ask, "at rest or in motion relative to what?"

The train is in motion relative to the landscape, and the tracks are at rest relative to the train. The conductor, who moves up the aisle collecting tickets, is in motion relative to the seated passengers, who are at rest relative to the train. However, the passengers are in motion relative to a boy sitting on a fence outside watching the train go by. The boy is either at rest or in motion, depending on what he is being compared to—the train, the fence, or the Sun. Absolute rest and absolute motion do not exist. Rest and motion are relative and depend on the frame of reference.

CHAPTER 3

SPACE AND TIME

Until the twentieth century, space and time were assumed to be separate. Space existed and time flowed, much like the water in a river, rushing downstream from the present into the future.

Newton thought space was fixed and unchanging, and that it served as a yardstick against which all motion could be absolutely measured. He also believed that there was an absolute, unchanging time. He might have imagined a great cosmic clock, ticking off the seconds from the moment the universe was created until the moment of its death. This, after all, seemed to make sense.

Einstein, however, showed that time, like motion, is relative and depends on the observer. He said that time and space are joined as a single entity, which he called space-time. Instead of imagining the universe in three dimensions, Einstein saw it as having four—length, width, depth, and time.

THE GEOMETRY OF SPACE-TIME

Einstein's special theory of relativity deals with inertial frames—frames of reference in which there is no acceleration or deceleration. A particle that is at rest remains at rest. A particle that is in motion neither speeds up nor slows down. This is Newton's first law. When Einstein spoke about space-time in his special theory, he saw it as flat. In mathematical terms, a flat universe is infinite.

In the general theory of relativity, which deals with gravity and acceleration, the geometry of space-time becomes curved. It may be spherical, like a ball, or warped, like a saddle. A curved universe cannot be infinite. A curved universe bends back on itself, so it is finite.

So is the geometry of space-time flat or curved? Based on recent measurements, it seems to be flat. It goes on and on forever, without boundaries. You may want to think about that the next time you look up at the sky on a clear, dark night.

THE FOUR DIMENSIONS

A three-dimensional coordinate system can be used to describe the position of an object in space. But suppose you want to set up a meeting with a friend. Specifying a location is not enough. You need a fourth dimension— time. It does no good to ask your friend to meet you at a restaurant for pizza unless you also say *when* you want to meet.

An event occurs in space and in time. We need a special coordinate system to show where and when an event takes place. We can't show four dimensions on a two-dimensional piece of paper, so we will use one coordinate, *x*, to represent the event's physical location. A second coordinate, *t*, will represent time. This is Einstein's space-time. The point representing the event moves horizontally along the *x* axis (through space) and vertically along the *t* axis (through time). As it moves, it describes a curved line called a world line. Everything that exists—from atoms to people to galaxies—describes a world line in the space-time coordinate system.

TIME IS RELATIVE TOO

Einstein's special theory shows that time is not at all what it appears to be. *When* an event occurs depends, not on some absolute clock, but on the observer. Two travelers moving relative to each other will not see time passing in quite the same way. Time is as relative as motion.

This does not mean that you can show up at school whenever you feel like it and blame your lateness on relativity. In our everyday lives, time is pretty much the same for all of us. Even though we are in motion relative to many things, we are not in relativistic motion—motion very near the speed of light. Time reveals its elastic nature only at very high velocities.

To understand the true nature of time, let's consider two transparent spaceships pointed in opposite directions. Ship 1 is traveling toward a distant star at nearly the speed of light. The ship's pilot is standing in its exact middle, holding a special double-ended flashlight. One end of the flashlight is sending a beam of light toward the front of the ship, and the other is sending a beam toward

Relative Time

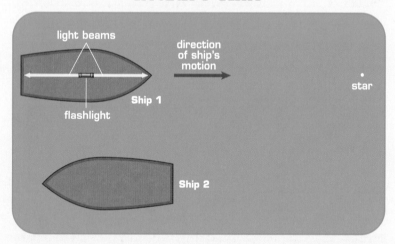

Ship 1 is moving toward a distant star at nearly the speed of light. Ship 2 is at rest relative to the same star.

the back of the ship. Light travels at a fixed speed. Since the front and back of the ship are exactly the same distance from the flashlight, the pilot sees the two beams hit the walls at the same instant.

As the pilot conducts this experiment, Ship 1 speeds past Ship 2, which is at rest relative to the star. The pilot of Ship 2 sees the front beam arrive after the back beam. Both pilots have made accurate measurements. So what's going on?

As Ship 1 shoots past, Pilot 2 sees the rear of Ship 1 moving toward the light waves emitted from the flashlight, while the front of Ship 1 moves away from them. The light waves, therefore, take longer to reach the front of the ship than to reach the rear of the ship.

Let's try the experiment again. This time, Pilot 2 holds the flashlight. Because she is at rest relative to the ship, she sees the beams hit both walls simultaneously. Because Ship 1 is traveling at a constant velocity toward the distant star, Ship 2 appears to be moving very fast in the opposite direction. So Pilot 1 sees the light taking longer to reach the front of Ship 2 than to reach the back. Pilot 2, who is at rest relative to Ship 2, sees no such difference.

The speed of light is the same for all observers. So the two pilots see all of the light beams moving at exactly the same speed. If the speed of light doesn't change, why do the pilots disagree about when the light beams hit the walls?

The answer lies in the nature of time itself. Time, like motion, is relative. When something happens depends on the observer. Because the pilots are in motion relative to each other, they observe the same events from two different frames of reference. In order for the speed of light to be the same in both frames, the time it takes the light beams to hit the front and back walls of the ships, as measured by the observers, must be different. This phenomenon is called time dilation.

TIME DILATION

Time dilation is one of the most fascinating aspects of special relativity. Simply put, an observer in motion relative to a clock sees the clock (and the time it is measuring) slow down. The greater the relative velocity, the slower the clock ticks.

Let's synchronize two identical clocks and put them in the two spaceships. Ship 1 is traveling toward a distant

Time Dilation

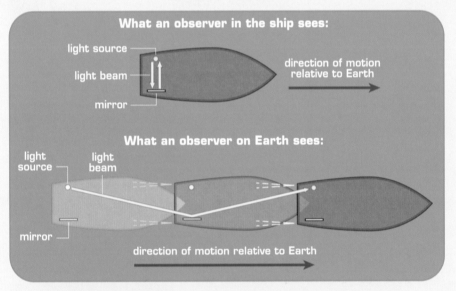

What an observer in the ship sees:

light source

light beam

direction of motion relative to Earth

mirror

What an observer on Earth sees:

light source

light beam

mirror

direction of motion relative to Earth

A ship is moving near the speed of light relative to
Earth. An observer in the ship sees the beam of light
travel across the ship and back. An observer on Earth
sees the light travel a much longer path. The speed of
light must be the same for both observers. Therefore,
from the point of view of the observer on Earth, time is
passing more slowly in the moving ship.

star, while Ship 2 is motionless relative to the star. This
is just for convenience. Ship 2 could be in motion relative to the star, and Ship 1 could be at rest relative to the
star. Or the two ships could be approaching each other at
relativistic velocities. When motion is uniform, it is
meaningless to ask who is at rest and who is in motion.

As Ship 1 shoots past Ship 2 at nearly the speed of
light, Pilot 2 notices that Ship 1's clock seems to be running slower than Ship 2's clock. Pilot 1 insists that his
clock is on time. Since the clocks have been synchronized and work perfectly, why is there a difference?

LINGERING MUONS

Cosmic rays are super-high-energy subatomic particles that move through space at nearly the speed of light. They rain down on Earth from all directions. To detect them, scientists use a device called a cloud chamber. The chamber is filled with a vapor, such as water vapor, and subjected to a magnetic field. When subatomic particles shoot through the vapor, they leave behind tracks that are visible as tiny droplets *(right)*. A particle's mass and charge determine what kind of track it makes. By examining the tracks, scientists can identify the particles that produced them.

In 1935, while conducting cosmic ray experiments at the top of Pike's Peak in Colorado, the physicist Carl Anderson noticed a kind of track he had never seen before. It was less curved than an electron's track but more curved than a proton's track. He had found traces of a previously unknown particle. Anderson called the new particle a mu meson, or muon for short.

Muons form high in the atmosphere. Two microseconds (two millionths of one second) later, they decay (break apart). The journey from the upper atmosphere to the top of Pike's Peak was simply too long for muons to have survived intact. What, then, were they doing there? They were offering proof that time dilation is a real phenomenon. Their relativistic velocity extended their life spans. Because they were in existence longer, they had time to travel farther.

Because Ship 1 is traveling at such an enormous velocity, its time is slowing down. Pilot 1 doesn't notice, however. He is inside the ship and therefore is also affected by time dilation. His clock appears to be ticking normally, and it is—within his frame of reference. But when Pilot 1 looks at Ship 2's clock, he sees it running slow relative to his clock. Pilot 2 sees her clock running perfectly, and it is—within her frame of reference.

Who is right? Once again, the answer is that both pilots are right. To Pilot 1, Pilot 2's clock is slow. To Pilot 2, Pilot 1's clock is slow. But to both of them, their own clocks appear to be ticking normally.

THE TWIN PARADOX

Relativistic speeds slow down clocks. What does that mean for astronauts traveling close to the speed of light? Would they live longer than they would have back on Earth?

Let's consider two twenty-year-old identical twins, Mike and Ike. Mike and Ike are the same in every way except one: Mike is adventurous and Ike is not. Mike decides to head off across the galaxy in a spaceship, while Ike remains on Earth. When Mike's rocket reaches a distant star, he will turn around and come home. Relative to Earth, Mike's rocket will travel at a constant speed of 150,000 miles (240,000 km) per second—80 percent of the speed of light.

Mike and Ike synchronize their clocks, and Mike takes off. As Ike waits back on Earth, the clock in his kitchen shows time passing in the usual way. Time also passes in the usual way for Mike, as measured by the clock aboard his spaceship. But because Mike is moving at a relativistic speed, time inside his ship has slowed down. At 80 percent of the speed of light, only three seconds pass in Mike's spaceship for every five seconds that pass on Earth.

The longer Mike is gone, the more the difference adds up. If Mike returns home after thirty ship years of traveling, he will find that fifty years have passed on Earth. His brother will be seventy, but Mike will be only fifty years old. What will happen if Mike decides to continue traveling at 80 percent of the speed of light and never come home? It won't do him any good. In order for Mike to reap the age benefits of his relativistic flight, he has to return to Earth.

Mike can't tell that time has slowed down within his ship, because everything in the ship is in relativistic motion together. Everything is aging more slowly—not just Mike. Time is consistent within Mike's frame of reference. It is only when Mike leaves this frame of reference and moves into another frame of reference—Earth—that the time difference manifests itself.

Mike is in uniform motion relative to Earth. That means when Mike and Ike look at each other through their telescopes, they are unable to tell who is in motion. When Mike looks at Ike's kitchen clock, he sees it running slower than his shipboard clock. That means Ike, back on Earth, must be aging more slowly than he is!

How can Mike simultaneously age more slowly and less slowly than his brother? He can't. That's why scientists call this scenario the twin paradox.

Mike and Ike have taken different paths through space-time. Because one twin is in relativistic motion and the other is not, their world lines are different. Mike's world line, from Earth to the star to Earth, winds up being shorter than Ike's world line, even though Ike didn't go anywhere. Mike's relativistic speed provided him with a kind of shortcut.

Some years ago, scientists tested the theory of time dilation. They used a pair of atomic clocks—extremely accurate clocks that are regulated by the vibrations of an atom or molecule. One clock flew around the world on a jet, while its synchronized twin stayed home. A jet can't go anywhere near relativistic speeds, but the results of the experiment nevertheless showed that speed does affect the passage of time. When the clocks were reunited at journey's end, the traveling clock had recorded a shorter elapsed time than its stay-at-home twin.

CHAPTER 4

LIGHT, SPEED, MASS, AND ENERGY

Special relativity makes a number of astonishing predictions about how matter behaves at relativistic speeds. At nearly the speed of light, the universe reveals a side of itself we never see.

THE DOPPLER EFFECT

Travelers in a spaceship moving close to the speed of light will see an extraordinary view through their ship's windows. This is due to a phenomenon known as the Doppler effect, or the Doppler shift. It is named for the Austrian physicist Christian Doppler, who first described it with respect to sound waves.

Imagine standing at a railroad crossing. A westbound train is barreling down the track at a constant speed, its whistle blaring. You notice that the whistle's pitch drops dramatically at the moment that the train passes you. Train whistles produce only one pitch. The engineer who is on the train with the whistle hears no such change in pitch.

Why do we hear a change, while the engineer doesn't?

Let's put a cow on the tracks. The engineer spots the cow and manages to stop the train several feet away. He blasts the whistle. The whistle sends out sound waves in all directions. It's somewhat like the ripples made by a stone when it is dropped into a pond. The ripples spread out evenly from the source. We, who are stationary relative to the train, hear one continuous pitch as the sound waves reach our ears.

But consider what happens when the train is moving. As the whistle emits sound waves, the train moves closer and closer to us. So the waves reach our ears more frequently than they would if the train and its whistle were standing still.

The Doppler Effect

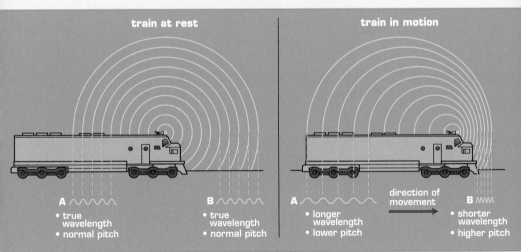

train at rest

train in motion

A
- true wavelength
- normal pitch

B
- true wavelength
- normal pitch

A
- longer wavelength
- lower pitch

direction of movement

B
- shorter wavelength
- higher pitch

Sound waves from a moving train's whistle are compressed in front of the train and stretched out behind it.

A sound's pitch is determined by the frequency of the sound waves. Low-frequency waves are farther apart, while high-frequency waves are more bunched together. When the train is heading toward us, the sound waves from the whistle are bunched together in the direction of motion. When the train passes us, the waves become stretched out, and the pitch falls.

Electromagnetic radiation is also subject to the Doppler shift. Each particular kind of radiation has a different range of wavelengths. Radio waves are the longest, and gamma rays are the shortest. Visible light falls somewhere in the middle.

When sunlight shines through droplets of water in the atmosphere after a rainstorm, we often see a rainbow. A rainbow is what happens when visible light is split into its component colors: red, orange, yellow, green, blue, indigo, and violet. These colors have their own distinctive wave-

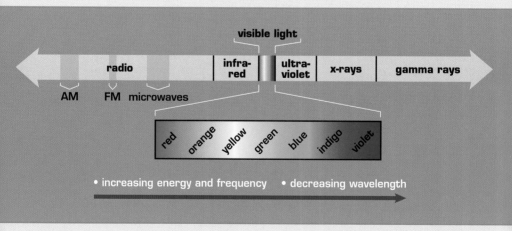

The Electromagnetic Spectrum

visible light

radio | infra-red | ultra-violet | x-rays | gamma rays

AM   FM microwaves

red   orange   yellow   green   blue   indigo   violet

• increasing energy and frequency   • decreasing wavelength

lengths, with red being the longest and violet the shortest. After violet come the ultraviolet wavelengths, which we can't see. Nor can we see infrared wavelengths, which are longer than red.

If you look up at the sky at night, you may at first think that all stars are silver. But if you stare at them long enough, you will realize that's not the case. Stars radiate light of different wavelengths, depending on their temperature, so they come in all colors. And because stars also radiate in wavelengths that are outside the visible light range, some stars can't be seen by the naked eye. Astronomers who study these stars use telescopes that are designed to detect wavelengths outside the visible range.

A photograph of the night sky reveals stars in a variety of colors.

As an Earthly stargazer, the star colors you see are like the steady, unmoving blast from the stationary train whistle in the previous example. Even though Earth and stars are in motion relative to one another, the motion is not fast enough to produce a Doppler shift. The Doppler shift becomes noticeable only when the source is approaching or receding at high speed relative to the observer.

Let's hop aboard a spaceship and find out how relativistic speeds affect what we see when we look at the stars. Our ship has two viewing ports: one in the front and one in the back. The ship is moving toward the stars that we can see through the forward port and away from the stars that we can see through the aft port. We have a star chart that shows the colors of the stars as seen from Earth.

As our ship begins to accelerate, you compare the colors of the stars you see through the fore and aft ports with the colors indicated on the star chart. The colors match. Red stars look red, blue stars look blue, and yellow stars, like the Sun, look yellow. But as we increase our speed, something remarkable happens. The colors begin to change as the electromagnetic radiation emitted by the stars becomes Doppler shifted. The colors of the stars in the direction of the ship's motion appear to shift to shorter wavelengths. Scientists call this blueshift, because colors are shifted toward the blue end of the spectrum. Red stars become orange, yellow stars become green, and blue stars become indigo—very deep blue, almost purple. But perhaps the strangest shift occurs in stars that are radiating in the invisible wavelengths. Stars that radiate in the infrared pop into view as red stars because their longer wavelengths now appear shortened. Violet stars cross the line into the ultraviolet and wink out.

Behind us, it's just the opposite. We are rushing away from those stars, so their wavelengths appear longer—they are redshifted. Stars radiating in the ultraviolet suddenly appear as violet stars. Blue stars appear green, green stars appear yellow, and yellow stars become orange. Thanks to the Doppler shift, our acceleration to relativistic speeds causes the entire universe to ripple like a cosmic light show.

THE ABERRATION OF LIGHT

Relativistic speeds change more than star colors. The scenery changes as well. When you look through the forward window before we begin to increase our speed, you can easily make out the constellations Cassiopeia and the Big Dipper. But at 40 percent of the speed of light, the stars seem to be drawing closer together. Stars that were out of our field of vision when we started our trip are beginning to crawl into view. Soon we can see the constellation Orion. At 70 percent of the speed of light, the stars are bunching up even more, and Orion has climbed higher into our field of view.

This phenomenon is called aberration of light, from the Latin word *aberrare*, meaning "to go astray." It occurs because our spaceship is in motion relative to the stars. We see the same thing happening back on Earth as our planet travels around the Sun, but the effect is not nearly as pronounced because Earth is moving much more slowly than our ship.

If you have ever walked through a driving rain with an umbrella, you've experienced something similar. If the raindrops are falling straight down and you are standing still, you can hold the umbrella directly over your head and stay dry. But if you begin to walk, you have to tilt the umbrella because you are moving into the falling rain. The faster you walk, the more you have to tilt the umbrella to stay dry. In other words, when you are moving with respect to the vertically falling raindrops, the raindrops seem to be coming at you at a different angle.

Inside our spaceship, we are seeing a displacement of starlight. The stars that make up the constellation Orion are not actually in our field of view. They only seem to be because of our tremendous speed. Their light has been displaced, or shifted. Light from all the other stars is also being displaced. The faster we go—as if we were running through the rain—the greater the displacement. That is why the stars look as if they are crushing together.

If we look out the front window at 99.98 percent of the speed of light, the familiar constellations are gone. All the stars seem to have gathered together in a tight cluster. The stars behind the ship—those we are moving away from—have had their light displaced so much that they appear to be in front of us. If we turn around and look out the aft window, we see nothing but darkness.

A journey in a spaceship traveling close to the speed of light would be strange indeed. But what would happen if we traveled at exactly the speed of light? What would we see then?

THE FORBIDDEN VELOCITY

Special relativity says that nothing can accelerate to the speed of light. Objects can come very close, as do some subatomic particles. Scientists have theorized that there may be some particles that always travel faster than light. But accelerating to exactly the speed of light is forbidden. Why?

FASTER THAN LIGHT In 1931 the Austrian American physicist Wolfgang Pauli proposed the existence of a particle that his colleague Enrico Fermi would later call the neutrino. Pauli speculated that neutrinos would be massless and travel at the speed of light, just like photons. After a good deal of searching, neutrinos were finally detected in 1956. That's when physicists began to wonder if there could be a class of particles that only travel *faster* than light. They called these particles tachyons, or "rapid units."

Special relativity says that nothing can accelerate to the speed of light, but the theory does not forbid anything from starting out at the speed of light—as neutrinos do—or from moving faster than light. This means tachyons would have to always travel faster than light. They could never slow down to cross the light speed barrier from the opposite side.

The speed of light, then, is a fence that separates the tardyons (slower-than-light particles) from the tachyons. It would be impossible to show tachyons in a space-time diagram. They would not exist in a past or a future but somewhere else, outside of space-time.

Tachyons have not yet been detected. If they do exist, the universe may prove to be even stranger than it already seems.

Some years before Einstein published his papers on special relativity, Hendrik Lorentz made some extraordinary predictions about fast-moving objects. He showed mathematically that moving bodies approaching the speed of light contract (shorten in length) in the direction of motion. This is called the Lorentz-Fitzgerald contraction because the Irish physicist George Fitzgerald had come to the same conclusion (unbeknownst to Lorentz) a few years earlier.

Both Lorentz and Fitzgerald showed mathematically that if an electron were moving at 90 percent of the speed of light, it would effectively have twice as much mass as if it were at rest. As the particle drew closer and closer to the speed of light, it would become more and more massive. At exactly the speed of light, the particle would have infinite mass!

Mass has nothing to do with an object's size, shape, or charge. A sumo wrestler might be 6 feet (2 meters) tall and nearly as wide and be less massive than a wrecking ball with a diameter of 2 feet (0.6 m). A star can be the size of a marble and be more massive than either of them. Mass is determined by how much matter makes up the sumo wrestler, the wrecking ball, or the star. The more mass an object has, the greater the force that is needed to move it.

Let's board our spaceship again and see what happens when we attempt to reach the speed of light. When the spaceship is on the launchpad, it is at rest relative to the ground. Because of its inertia, it takes a good bit of fuel to supply the thrust to get it moving. But once it has achieved a velocity of about 7 miles (11 km) per second, it breaks free of Earth's gravity. This speed is called Earth's escape velocity.

We head off through the solar system. As we continue to accelerate, we notice that we have to use more and more fuel to increase our velocity. Why?

The Fitzgerald-Lorentz contraction predicts that the faster we go, the more massive our ship becomes. And the more massive our ship, the more fuel we have to use to continue to accelerate it. We will never be able to achieve the speed of light because at that exact velocity our ship would become infinitely massive.

Then how are photons able to travel at the speed of light? Photons have no mass. Because they have no mass,

PARTICLE ACCELERATORS A particle accelerator is a device that uses an intense magnetic field to accelerate charged particles to speeds very close to that of light. There are two basic designs. A linear accelerator propels particles in a straight line, while a cyclic accelerator sends them around a circular course.

As relativity predicts, the mass of the accelerated particles increases as their velocity increases. An electron orbiting a cyclic accelerator at maximum energy has a mass much greater than its rest mass. As the electron accelerates and its mass increases, the more energy it takes to keep the particle moving. This is similar to a spaceship using more and more fuel as it accelerates closer and closer to the speed of light. The behavior of particles in accelerators confirms the special theory of relativity.

they are able to travel at a speed the rest of us can only dream about.

MASS-ENERGY EQUIVALENCE

In the fifth paper he published in 1905, Einstein turned his attention to mass and energy. The paper was titled, "Does the Inertia of a Body Depend Upon Its Energy Content?" Nobody before Einstein had asked such a question. The answer would turn out to be one of Einstein's most famous and extraordinary statements about matter, energy, light, and motion. It would also ultimately change the world. What Einstein realized was that energy and matter are interchangeable.

In our everyday lives, we frequently use the word *energy* when talking about ourselves or other people. We say that some people have a lot of energy, but others never seem to have enough energy to finish a job. In physics, energy is defined as the capacity to do work. Besides the things we usually consider to be work, it includes fun things like hitting a tennis ball, swimming, or skiing. Work is also done when a book falls off a table or when the wind turns the blades of a windmill.

If you pick up a cinder block, you are doing work. The amount of work done equals the mass of the block times the distance you lift it. The heavier the weight and the greater the distance it is moved, the more work is done.

Many things possess energy and have the capacity to do work. This stored energy is called potential energy. It is energy waiting to be released, like that of a marble poised at the top of a *U*-shaped track. Kinetic energy is another kind of energy. It is the energy of motion—the motion of the marble rolling down the track. When the marble reaches the bottom of the track, the kinetic energy of its downward motion moves it up the other side of the track.

If you push something, such as a marble, you supply it with kinetic energy, the energy of motion. And when an object is set in motion, its mass increases. The increase in the marble's mass is extremely small, of course, because the marble is not moving very fast. The marble's additional mass comes from its motion.

The principle of conservation of energy says that energy can neither be created nor destroyed. It can only be converted from one form to another. We see examples of energy transformations every day. When a car burns gasoline, chemical energy is converted to heat energy, which lifts the pistons in the cylinders and becomes mechanical energy.

If matter is a form of energy and one form of energy can be converted into another form, it follows that matter can be converted into energy. Einstein's famous equation $E=mc^2$ describes how much energy is locked up in a given amount of matter. In this equation, E represents the amount of energy, m stands for the matter's mass, and c stands for the speed of light. So the amount of energy locked up in a given amount of matter is equal to its mass multiplied by the speed of light squared.

Plugging numbers into Einstein's equation, we learn that 2.2 pounds (1 kilogram) of matter is equivalent to an astonishing 25 billion kilowatt-hours of energy. This is enough energy to lift something weighing 1.1 billion tons (1 billion metric tons) to a height of 5.9 miles (9.5 km)—higher than Mount Everest! The tough part is figuring out how to release this energy.

Mass-energy conversion experiments began in the 1930s. They culminated at Alamogordo, New Mexico, on July 16, 1945, when physicists succeeded in releasing the energy locked within the atom. The explosion of the world's first atomic bomb unquestionably confirmed Einstein's theory that mass could be converted into energy.

The mass-energy paper rounded out Einstein's special theory of relativity. In less than one year, Einstein had completely overturned the prevailing view of the universe.

CHAIN REACTION Nuclear fission is often called splitting the atom, but it is actually an atom's nucleus that is split. In the process, some of the mass of the nucleus is converted to energy. Elements such as uranium and plutonium are used for nuclear fission because their atoms have very heavy nucleii—each nucleus consists of large numbers of protons and neutrons.

An atomic nucleus is split by hitting it with a high-speed subatomic particle. The collision releases an enormous amount of energy. High-energy photons called gamma rays are produced, and neutrons from the nucleus fly off at high speeds. The neutrons blast apart other atomic nucleii, causing a chain reaction. Once a chain reaction begins, it becomes self-sustaining. Unless it is controlled, the result is an atomic explosion.

CHAPTER 5

GENERAL RELATIVITY

In his special theory of relativity, Einstein had overturned the old belief that space was fixed and unchanging. He had shown that uniform motion was relative to the observer. But a universe without a universal fixed frame of reference should suggest that all motion is relative. Therefore, acceleration and deceleration should be relative too. But a simple train ride shows that this is not true. Even with your eyes closed, you know when the train starts moving, when it slows down, and when it pulls into a station and stops.

Einstein did not believe that the rules should change depending on the situation. So for more than a decade, he pondered and wrestled with the physics. He revisited Newton's law of universal gravitation and Galileo's law of falling bodies. Finally, by looking at gravity in an entirely new way, it all came together. In 1916 Einstein published a paper titled "The Foundation of the General Theory of Relativity."

THE PRINCIPLE OF EQUIVALENCE

In his famous (but perhaps entirely legendary) experiment, Galileo dropped two cannonballs, one ten times as heavy as the other, from the Leaning Tower of Pisa. Despite their difference in weight, the cannonballs hit the ground at the same time because both had been accelerated at the same rate—32 feet (9.8 m) per second per second, which is written as 32 ft/s^2 (9.8 m/s^2). What had caused the acceleration? It had to be Earth's gravity. This led Einstein to conclude that an observer cannot distinguish between the effects of gravity and the effects of a constant rate of acceleration. He called this idea the principle of equivalence.

Einstein imagined two identical laboratories, one on Earth, and the other in space, far from any object whose gravity might affect it. He imagined that an angel picked up the sky laboratory and flew off with it, giving it the same acceleration experienced by falling objects on Earth. Essentially, Einstein was creating gravity for his space laboratory by accelerating it. This would cause conditions inside the space laboratory to be the same as those inside the Earth laboratory. If a scientist in the space laboratory and a scientist in the Earth laboratory simultaneously dropped two identical flasks from the same height, the flasks would hit their respective floors at the same time. Uniform acceleration, therefore, is equivalent to gravity.

Then it occurred to Einstein that if a person were falling freely, he would not be able to feel his own weight. Imagine that an elevator is stopped at the tenth floor of an office building. You are inside, holding a ball. If you drop the ball, it falls to the floor of the elevator because it is subject to Earth's gravity. As long as the elevator is stationary, conditions inside it are the same as conditions in the lobby, the offices, and the street outside.

The doors now close, and the elevator starts to descend. On the way down, the cable snaps, and the elevator plummets down the shaft. You are in what is called free fall. You are under the influence of only one force—gravity.

Instead of panicking, you quickly drop the ball again to see what will happen. The elevator is accelerating at 32 ft/s^2, but so is the ball. To you, an observer within the falling elevator, the ball seems to be at rest relative to the elevator. Instead of dropping to the floor, it appears to float. Since the ball is accelerating with the elevator, it can't overtake it and fall relative to it. You have just discovered that free fall cancels the effects of gravity. This is why the astronauts in an orbiting space station do not come to rest but float weightlessly.

The principle of equivalence tells us that constant acceleration is not absolute but relative. Imagine you are inside a small, windowless room that has an overhead sprinkler. The sprinkler goes on, and water begins to fall. The theory of general relativity says you cannot assume that the room is at rest and Earth's gravity is responsible for the downward motion of the drops. The room could, instead, be accelerating upward. Furthermore, there is no experiment you could do inside the room to decide whether constant acceleration or uniform gravity is responsible. They produce the same effects.

MASS AND TIME

Let's return to Einstein's space laboratory. We find that a scientist has attached a bowling ball to a spring. As the laboratory is accelerated at 32 ft/s², the spring stretches out, indicating the bowling ball's resistance to being set into motion. This resistance is called its inertial mass. An object with a large inertial mass is harder to set in motion than one with a small inertial mass. Meanwhile, in the Earth laboratory, a scientist has set up the same experiment. Here, there is no acceleration. There is, however, gravity, so we say the bowling ball has gravitational mass. According to the principle of equivalence, both springs should stretch the same amount. And that is what happens. Inertial mass is equivalent to gravitational mass.

Next, the scientists add the same amount of kinetic energy to each bowling ball. According to Einstein's mass-energy equation, $E=mc^2$, adding energy increases an object's mass. The equally extended springs show that the mass increase for the accelerated bowling ball is indeed the same as the mass increase for the gravitationally affected bowling ball.

What happens to time? In the space lab, a scientist has two clocks that send out a pulse of light each time they tick. She synchronizes the clocks with her wristwatch. Then she places one of the clocks on the floor and attaches the other to the ceiling. As her lab is being accelerated at 32 ft/s², the scientist looks down at the clock on the floor. It appears to be ticking more slowly than her wristwatch. The clock on the ceiling, on the other hand, appears to be ticking more quickly than her watch. Why?

In the accelerating lab, the scientist's watch is moving away from the floor clock's light pulses and toward the ceiling clock's pulses. The light pulses of both clocks travel at a constant speed. Pulses from the floor have farther to go than pulses from the ceiling. Therefore, the floor clock seems to run slow and the ceiling clock seems to run fast.

Back on Earth, a scientist conducts the same experiment. His lab isn't accelerating, but it is affected by Earth's gravity. The principle of equivalence tells us that he will see exactly the same thing as the scientist in the space lab.

Einstein predicted that two clocks placed at different distances from a gravitational source (such as Earth) will appear to run at different rates. An observer at the top of Mount Everest will see a clock at sea level running slower than a clock on the ground next to her. Conversely, if she returns to sea level, she will see the clock on Mount Everest running faster than the clock at her feet. How much faster or slower depends on the strength of the gravitational field.

GRAVITATION AND LIGHT

Suppose we put some glowing atoms on the ceiling and the floor of the accelerating space laboratory. Our scientist looks down at the atoms on the floor. The frequency of light they emit seems to be lower—the glowing atoms are redshifted. The frequency emitted by the atoms on the ceiling seems to have increased. The glow of these atoms is blueshifted.

The Sun is a very massive body with a powerful gravitational field. Einstein's principle of equivalence predicted that light from glowing atoms on the surface of the Sun should appear to have lower frequencies than light emitted by similarly glowing atoms coming from a source with a weaker gravitational field. In other words, the Sun's light should be redshifted by the Sun's gravitational field. Experiments eventually confirmed this prediction. In the 1960s, experimenters at Harvard University detected a tiny gravitational redshift between the top and the bottom of an 877-foot-tall (267 m) tower on campus.

The general theory of relativity tells us that gravitation and acceleration bend light rays. You might imagine a scientist shining a very powerful flashlight horizontally across our Earth laboratory. The light coming from the flashlight tries to travel in a straight line (like the Moon trying to obey Newton's first law), but because it is affected by Earth's gravity, it is bent downward.

This bending of light by gravity was confirmed in an experiment during a total eclipse of the Sun in 1919. The British astronomer Arthur Eddington led an expedition to the island of Principe, off the coast of Guinea in western Africa. During totality—the few minutes when the lunar

Arthur Eddington

This photograph shows the instruments used by Arthur Eddington and his expedition to observe the 1919 solar eclipse.

disk blocks out nearly all of the Sun's light—astronomers would be able to see the light from other stars passing close to the Sun. Usually this is impossible, due to the Sun's brightness. If Einstein's theory was correct, the Sun's gravitational field would bend the starlight.

The relativity experiment required two measurements of the target star's location relative to other stars. One was taken during the eclipse. The other was taken six months later, when the star was directly opposite the Sun. At that time, the starlight was not passing close to the Sun, so it was not affected by the Sun's gravity. The experiment unfolded as the general theory predicted. The bent light path revealed itself when the two measurements were compared.

BLACK HOLES

Stars shine by converting hydrogen and helium into heavier elements through nuclear fusion reactions. The outward pressure created by nuclear fusion counteracts the inward pull of gravity and prevents the star from collapsing. But once a star's core has been converted to carbon, the reactions shut down. Gravity takes over, and the star collapses under its own weight.

Low-mass stars become dense white dwarfs with diameters of only a few thousand miles. A white dwarf's thin, gaseous atmosphere slowly leaks away into space. Eventually, the star cools and becomes a black dwarf.

General relativity predicts a different fate for stars with masses more than eight times the mass of the Sun. When such a star has exhausted its fuel, it blows off its outer layers in a huge supernova explosion. All that is left is a small, dense core. The core begins to contract. As its atoms are packed tighter and tighter, the core's gravity increases, causing it to collapse even faster. It becomes still smaller and denser, until it reaches the critical size at which its escape velocity exceeds the speed of light. Then the star winks out and becomes a black hole.

Nothing—not even light—can escape a black hole. Therefore it cannot be observed directly. Its existence can only be inferred by its effects on the space around it. In 1996 astronomers found strong evidence for a massive black hole at the center of the Milky Way galaxy. It's possible that black holes exist at the center of most other galaxies too.

Gravity Bends Starlight

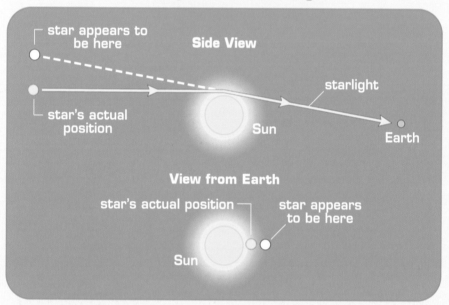

Normally light travels in straight lines, but it can be bent by the curvature of space near a large mass such as the Sun.

EINSTEIN'S THEORY OF GRAVITY

Newton understood that the gravitational attraction between two objects depends on the distance between them and their masses. This relationship is called Newton's law of universal gravitation. Gravity apparently did not need a medium to carry it. It seemed to operate instantaneously, and it decreased as the distance between the objects increased. Newton assumed that gravity was a force, but he had no idea what the nature of that force could be.

Newton's world had three dimensions, in which space extended for an infinite distance. Objects existed in space without affecting space in any way. When Einstein joined

John Wheeler

space and time into four-dimensional space-time, a new geometry of the universe was needed. Einstein realized that gravity was not a force but the interaction of objects in space-time. The American physicist John Wheeler summed it up when he said, "Space tells matter how to move, and matter tells space how to curve."

Suppose we set up a trampoline and place a large boulder in the center. The boulder makes a considerable dent in the fabric of the trampoline. We will call this dent the boulder's gravitational field. When we place a baseball on the trampoline, it, too, makes a dent, but one that is much smaller. The reason is that the boulder has much more mass than the baseball.

If we place the baseball close enough to the boulder, the baseball will roll toward the boulder. The baseball is attracted by the boulder's gravitational field. But the baseball also has a gravitational field. Why wasn't the boulder attracted to the baseball? It was, but the boulder had more influence on the baseball than the baseball had on the boulder.

In four-dimensional space-time, all objects make gravitational "dents" around themselves. They distort space-time. Gravity does not create the curvature; it is the curvature. Matter, in turn, is affected by this curvature. When an object approaches another object's gravitational field, its motion through space cannot continue in a straight line. Its path must bend. If you were to roll a marble in a straight line across the trampoline, its path would curve if it came close enough to the dents made by the boulder or the baseball. As the 1919 eclipse experiment showed, even photons have to follow the curvature of space-time.

GRAVITATIONAL WAVES

According to Einstein's general relativity theory, gravitational waves are produced when a very massive body is accelerated through space-time. You might imagine them as ripples like those produced when a stone is dropped into a lake, except that gravitational waves travel at the speed of light. The theory predicts that when these waves encounter an object, such as a planet, they should make it vibrate slightly.

Gravitational waves have not yet been detected, but astronomers are studying a double star system that may be producing them. In a double star system, two stars are close enough to be affected by each other's gravitational field. One of the stars in this particular system is a pulsar—a small, very dense, rapidly rotating star that emits bursts of radiation in regularly occurring pulses. Measurements have shown that the pulsar's orbit is slowing, corresponding to a decrease in the energy of its orbital motion. The generation of gravity waves may be responsible for this energy decrease.

CHAPTER 6

EINSTEIN'S LEGACY

When Einstein developed his theories of relativity, astronomers knew very little about the universe. They hadn't even discovered the planet Pluto yet. But by giving scientists a new way to consider matter and energy, gravity, and a four-dimensional space-time, special and general relativity sent them off into previously unthought-of directions. Astronomers and physicists began to apply the principles of relativity to their own particular areas of investigation. The result was an explosion of unique ideas and theories. The impact of relativity on physics and astronomy has been nothing short of enormous.

AN EXPANDING UNIVERSE

Months after Einstein published his general theory of relativity in 1916, the Dutch mathematician and astronomer Willem de Sitter published a series of papers describing general relativity's applications in astronomy. Einstein had envisioned the size of the universe as unchanging.

De Sitter did not agree. He pointed out that general relativity suggested that the universe was actually expanding in size. The Russian mathematician Alexander Friedmann and the Belgian astronomer Georges Lemaître also believed that the universe was expanding.

In 1927 Lemaître took this idea one step further. He suggested that if the universe is expanding, there must have been a time when everything was packed together in a "cosmic egg." Lemaître's cosmic egg contained all the matter in the universe, crushed down to a small, dense point. The cosmic egg exploded in what is known as the Big Bang, and the expansion of the universe began. We see the results of the explosion, billions of years after it occurred. At first, nobody paid much attention to Lemaître's idea.

In 1912 the American astronomer Vesto Slipher had discovered that all but a few of the galaxies we can see are receding from us. In 1929 another American astronomer, Edwin Hubble, analyzed this recessional speed. He found that the farther a galaxy is from us, the faster it is receding. This, he said, could be explained only if the universe is expanding.

Because he believed that the size of the universe was unchanging, Einstein had added a term called a cosmological constant to his equations. The term was essentially a force counteracting gravity. It would keep the universe in balance, preventing the galaxies from attracting one another and eventually collapsing in on themselves. But when Friedmann designed a mathematical model of an expanding universe, it showed that the

THE BIG CRUNCH Ever since the big bang theory was proposed, astronomers have speculated on the fate of the universe. Will it expand forever, as the cosmic jerk seems to suggest? Or will gravity eventually cause everything to collapse in a "Big Crunch"?

In the Big Crunch scenario, the expansion of the universe that began with the Big Bang eventually slows down. The universe experiences a contraction as space itself is drawn together. The closer together the galaxies get, the more they are affected by one another's gravity and the faster they contract. Like a movie running in reverse, the universe collapses into a point of infinite density and zero diameter, ending the way it began.

cosmological constant was unnecessary to account for the universe's expansion. Einstein later embraced the concept of an expanding universe, saying that the cosmological constant had been his greatest error. But was it?

American astrophysicist Adam Riess says that five billion years ago the universe experienced a "cosmic jerk." At that point, the expansion of the universe, which had been slowing due to gravity, began speeding up. Astronomers speculate that the cosmic jerk was caused by a hypothetical form of energy they call dark energy.

DARK MATTER

As part of general relativity, Einstein developed a set of field equations—equations that describe how fundamental forces interact with matter. The field equations make it possible to figure out how much curvature an object's gravitational field will create in space-time.

In 1917 Willem de Sitter came up with an interesting solution to Einstein's field equations. His solution showed that a massless universe would expand. In 1932 de Sitter and Einstein together came up with the Einstein-de Sitter model of the universe. In this model, most the matter in the universe is invisible "dark matter." It can't be observed directly because it doesn't emit electromagnetic radiation, but its gravitational effects on visible matter can be measured. Dark matter may account for more than 90 percent of the mass in the universe.

In 1997, while studying photographs taken by the Hubble Space Telescope, astronomers at Bell Labs noticed that light coming from one cluster of galaxies was bent by the gravity of another, nearer cluster. General relativity predicts that this should happen. But the astronomers found that the light was bent more than it should, based on the amount of matter visible in the nearer cluster. They estimated the mass of the foreground cluster to be 250 times the visible mass. Could dark matter account for the difference? No one knows yet, but the scientists' findings are certainly intriguing.

EINSTEIN'S DREAM AND THE NEW PHYSICS

Einstein once said, "God does not play dice with the universe." He deeply believed that the universe was ultimately governed by simple rules that described the behavior of everything, from the movement of atoms to the evolution of galaxies. Nothing was left to chance, nothing was uncertain.

Einstein's ultimate goal was a simple mathematical formula—a unified field theory—that would unite all the properties of matter and energy. It would describe the relationship of the four known forces: gravity, electromagnetism, and the strong and weak interactions that occur between subatomic particles. He labored, year after year, to find a mathematical relationship between electromagnetism and gravity.

Einstein's peers did not believe that a unified field theory was possible. They were saddened by what they considered Einstein's stubborn and futile quest. A new physics—quantum mechanics—was on the horizon. Quantum mechanics developed from Max Planck's quantum theory. It explains the mechanics of the universe—the behavior of particles, atoms, and molecules. While quantum mechanics could coexist with relativity, it seemed to show that a sort of dice game really was going on.

In the sixteenth century, the French astronomer and mathematician Pierre Simon LaPlace said that if, for one instant, you could know the position and velocity of every particle that exists, you could then calculate the entire history of the universe. But in 1927, the German physicist Werner Heisenberg discovered the uncertainty principle, which states that one can never be exactly sure of both the position and the velocity of a particle. The more accurate your measurement of one, the fuzzier the other gets. This uncertainty, most physicists believe, makes a unified field theory impossible.

AN UNSEEN UNIVERSE? Do other universes exist side by side with ours? Scientists know that empty space is not really empty. Particles appear and disappear, popping into existence and then popping back out. The theory of supersymmetry suggests that every known particle in the universe has an undetected counterpart.

Subatomic particles have a kind of built-in momentum called spin. Muons' spin is affected by magnetic fields. Scientists at Brookhaven National Laboratory on Long Island, New York, have found that they can't fully account for the way the magnetic field of their particle accelerator affects the spin of muons. The group believes that they may have caught a glimpse of a shadow universe filled with supersymmetrical particles.

Einstein failed in his attempt to develop a unified field theory, but his theories of relativity have stood the test of time. Every new theory that attempts to explain how the universe is put together, how it was created, how it is evolving, or how it might end is based on relativity. Joseph John Thomson, the discoverer of the electron, said of general relativity that it was "one of the greatest—perhaps the greatest—of achievements in the history of human thought."

Glossary

blueshift: the shifting of light toward the blue end of the spectrum because the light source is moving toward the observer

cosmological constant: a term Einstein added to his general relativity equations to counteract the force of gravity

dark energy: the unexplained energy in empty space that is causing the expansion of the universe to speed up

dark matter: matter in the universe that cannot be seen but can be detected by its gravitational effects on other bodies

Doppler effect: a change in the frequency of waves from a given source when the source and the observer are moving toward or away from each other; also known as Doppler shift

electromagnetic radiation: radiation consisting of electric and magnetic waves that travel at the speed of light. Examples include visible light, radio waves, gamma rays, and x-rays.

escape velocity: the speed at which a moving body breaks free of Earth's gravity; about 7 miles (11 km) per second

ether: the weightless, transparent, frictionless substance nineteenth-century scientists thought filled the universe, providing a medium through which light waves could travel

field equations: equations that describe how fundamental forces interact with matter

frame of reference: the perspective from which an observer perceives the universe

frequency: the number of times an electromagnetic wave oscillates in a given period of time

general theory of relativity: Einstein's theory that says that from a particular point of view, there is no difference between the effects produced by gravitation and acceleration

gravity: the force of attraction between objects with mass; the interaction of objects in space-time

inertia: a property of matter that causes it to resist changes in speed or direction

inertial frames: frames of reference in which there is no acceleration or deceleration

light-year: the distance light travels in one year

Lorentz-Fitzgerald contraction: moving bodies approaching the speed of light contract (shorten) in the direction of motion

magnetic field: area of magnetic force that exists around a magnet or a current-carrying conductor

mass-energy equivalence: the idea that mass can be turned into energy and energy can be changed into mass

oscillate: to move back and forth at a constant rate

photons: individual packets of electromagnetic energy

principle of equivalence: the idea that an observer cannot distinguish between the effects of gravity and the effects of a constant rate of acceleration

quantum (pl. quanta): the smallest unit of energy. A quantum of light is called a photon.

redshift: the shifting of light toward the red end of the spectrum because the light source is moving away from the observer

relativistic motion: motion at nearly the speed of light

space-time: the four-dimensional coordinate system (three dimensions of space and one of time) in which events take place

special theory of relativity: Einstein's theory that says the laws of physics are the same for observers moving at any speed and the speed of light is constant for all observers

time dilation: a stretching, or slowing, of time in a frame of reference moving past the observer at a speed approaching the speed of light

unified field theory: a theory that describes the four fundamental forces and the interactions between elementary particles within a single framework

wavelength: the distance from one peak of a wave to the next

world line: a path through four-dimensional space-time that represents a series of events

TIMELINE

1543 Nicolaus Copernicus publishes a Sun-centered model of the solar system.

1638 Galileo suggests that light appears to travel instantaneously because it moves so incredibly fast.

1675 Olaus Roemer proves that the speed of light is finite.

1687 Isaac Newton publishes laws of gravitation and motion.

1690 The first comprehensive theory of light was presented in 1690 by the Dutch physicist Christiaan Huygens, who said light behaves like waves.

1704 Isaac Newton proposes that light is composed of tiny particles.

1819 Hans Christian Oersted proves a connection between electricity and magnetism.

1842 Christian Doppler describes the Doppler effect on light and sound waves.

1854 Bernhard Riemann creates a four-dimensional geometry that Einstein will later use to describe space-time.

1882 Albert Michelson measures the speed of light traveling through a vacuum at 186,320 miles (300,000 km) per second.

1887 Albert Michelson and Edward Morley conduct an experiment that provides the first strong evidence against the luminiferous ether theory.

1900 Max Planck suggests that energy consists of tiny particles called quanta.

1905 Albert Einstein publishes his special theory of relativity.

1912   Vesto Slipher discovers that many galaxies are moving away from us.

1916   Einstein publishes his general theory of relativity.

1919   Arthur Eddington proves that gravity bends light rays.

1927   Georges Lemaître proposes that the universe expanded from a "cosmic egg."

       Werner Heisenberg states his uncertainty principle.

1929   Edwin Hubble finds that the farther a galaxy is from us, the faster it is receding.

1931   Ernest Lawrence builds the Cyclotron, the world's first particle accelerator.

1932   Willem de Sitter and Einstein propose the Einstein-de Sitter model of the universe, in which most matter is invisible "dark matter."

1935   Carl Anderson finds traces of a previously unknown particle—the mu meson, or muon.

1945   At Alamogordo, New Mexico, mass is converted to energy in the first atomic bomb.

1956   Frederick Reines and Clyde Cowan prove the existence of neutrinos.

1996   Astronomers find evidence of a massive black hole at the center of the Milky Way galaxy.

1999   Lisa Randall and Raman Sundrum propose a five-dimensional theory of gravity that competes with Einstein's general theory of relativity.

2002   William Harter uses the Doppler shift to derive the basic equations of relativity.

2006   Charles R. Keeton and Arlie O. Petters develop mathematical tools to test the Randall-Sundrum model of the universe.

Biographies

Christian Doppler (1803–1853) Doppler was born in Salzburg, Austria. He studied mathematics and astronomy in Vienna. He was appointed professor of mathematics and physics at the Technical Institute of Prague in 1841. One year later, he published a paper on what has since been called the Doppler effect. His ideas helped pave the way for the idea that the universe is expanding and made the weather-tracking technology known as Doppler radar possible.

Arthur Eddington (1882–1944) Eddington was a British physicist. He was a child prodigy, displaying a remarkable aptitude for mathematics very early. He graduated first in his class from Cambridge University, and in 1914 he became director of the Cambridge Observatory. Soon after, World War I broke out. A practicing Quaker and conscientious objector, Eddington was excused from military service. Eddington theorized that the reason stars do not collapse in on themselves is that the pressure of the heat they radiate counterbalances the inward pull of gravity. Eddington was knighted in 1930.

Albert Einstein (1879–1955) Einstein was born in Ulm, Germany. He enjoyed math and reading but disliked lectures and tests. His undistinguished university record led him to a job as a clerk in the Swiss patent office. In 1905—his *annus mirabilis*, or "miracle year"—he explained Brownian motion (the random motion of small particles suspended in a gas or liquid) and the photoelectric effect (the ejection of electrons from a metal surface when light shines on it) and introduced his special theory of relativity. In 1916 he published his general theory of relativity. In 1921 he won the Nobel Prize in Physics for his work on the photoelectric effect. He emigrated to the United States in 1933, when Adolf Hitler came to power in Germany, and became a professor of theoretical physics at

Princeton University in New Jersey. He acted as an unofficial adviser to U.S. president Franklin D. Roosevelt on the threat of the atomic bomb and was offered the presidency of Israel. In his private life, however, he had simple, quiet tastes. His hobbies included music and sailing.

MICHAEL FARADAY (1791–1867) Faraday was a British physicist and chemist. He received only the barest elementary school education. He was, however, a great theorist, able to envision what he could not explain with mathematics. Faraday worked as an apprentice for the famous chemist Humphry Davy, who quickly grew jealous of his young assistant's talent. When Faraday was elected to the Royal Society of London in 1824, it was the envious Davy who cast the single dissenting vote. Faraday discovered the relationship between electricity and magnetism, experimentally showing that one produced the other. He also invented an electric motor and a generator.

ALEXANDER FRIEDMANN (1888–1925) Friedmann was born in Saint Petersburg, Russia. He came from a family of composers, but he chose to study meteorology instead of music. After fighting in World War I, he became interested in Einstein's theory of relativity. He worked out a mathematical model of an expanding universe that suggested that some sort of massive explosion—now called the Big Bang—must have occurred at the very beginning of time. Friedmann's life was tragically cut short by typhoid fever.

GALILEO GALILEI (1564–1642) Galileo was born in Pisa, Italy. He studied medicine briefly, but he was more interested in physics and mathematics. Galileo was fascinated by pendulums and studied their movement in detail. He formulated the law of falling bodies and wrote a book on the science of mechanics. He designed and built a "magnifying tube" with which he discovered four of Jupiter's moons, the phases of Venus, sunspots, and the rotation of the Sun. Students loved to hear him speak, but many of his ideas angered the authorities, especially the Church of Rome. Galileo was tried by the Inquisition in 1633 and sentenced to house arrest. His brilliant book *Dialogue Concerning the Two Chief World Systems* remained banned by the church for almost two hundred years.

WERNER HEISENBERG (1901–1976) Heisenberg was a German physicist. He is best known for his revolutionary uncertainty principle, which he introduced when he was just twenty-six years old. In 1932 he won the Nobel Prize in Physics for his work in quantum mechanics. Many scientists left Germany during World War II, but Heisenberg stayed and worked on the development of the atomic bomb. The Los Alamos team in the United States achieved success first, however. After the war, Heisenberg continued his research, traveled widely, and became the director of the Max Planck Institute for Physics in Munich, Germany.

JAMES CLERK MAXWELL (1831–1879) Maxwell was a Scottish physicist whose extraordinary mathematical ability was already evident in grade school. At the age of fifteen, he submitted an original scientific article to the Royal Society of Edinburgh. He graduated with a degree in mathematics from Cambridge University in 1854. Two years later, he became a professor at Marischal College in Aberdeen, Scotland. Maxwell brought Faraday's theories of electricity and magnetism into clear focus in four equations that have come to be known as Maxwell's equations. He also studied such varied subjects as color blindness, photography, and the rings of Saturn.

ALBERT MICHELSON (1852–1931) Michelson was born in Strelno, Prussia (in modern-day Poland). His family moved to the United States when he was two years old. Michelson joined the U.S. Naval Academy. After graduation and a two-year cruise in the West Indies, he became an instructor in physics and chemistry at the academy. Michelson excelled in optics. His passion was measuring the speed of light, something he worked on until his death. In 1907 he became the first American to receive the Nobel Prize in Physics.

ISAAC NEWTON (1642–1727) Newton was born in Lincolnshire, England, shortly after his father died. When he was three years old, his mother remarried and sent him to live with his grandparents. He wasn't much of a student, preferring instead to design and build mechanical devices. By the time he was in his mid-teens, his grandparents had taken him out of school. Newton's uncle, however, insisted that Newton be sent to Cambridge University. Newton graduated, but he hardly stood out among his classmates. He began his musings on gravity, acceleration, and light as the Great Plague ravaged London. When he was just twenty-six, he was appointed Lucasian Professor of Mathematics at Cambridge. Newtonian physics reigned supreme for nearly two hundred years, refined and adjusted only by Einstein's theory of relativity.

MAX PLANCK (1858–1947) Planck was a German physicist. He was especially interested in the properties of light. His investigations led him to suggest that energy is radiated in discrete units that he called quanta. This discovery defined the break between classical and modern physics. In 1918 Planck was awarded the Nobel Prize in Physics for his quantum theory. In contrast to his great professional success, Planck had a tragic personal life. In 1909 his wife of twenty-two years died. He lost a son in World War I, and two daughters died in childbirth. During World War II, his house was demolished by Allied bombing. Another son was accused of plotting against Adolf Hitler and was executed.

SELECTED BIBLIOGRAPHY

Asimov, Isaac. *Asimov's Biographical Encyclopedia of Science and Technology*. New York: Doubleday & Company, 1982.

Chaisson, Eric. *Relatively Speaking*. New York: W. W. Norton & Company, 1988.

Davies, Paul. *About Time*. New York: Simon & Schuster, 1995.

Halpern, Paul. *Time Journeys*. New York: McGraw-Hill, 1990.

Hawking, Stephen, and Roger Penrose. *The Nature of Space and Time*. Princeton, NJ: Princeton University Press, 1996.

Hoffmann, Banish. *Relativity and Its Roots*. New York: W. H. Freeman, 1983.

Lilley, Sam. *Discovering Relativity for Yourself*. London: Cambridge University Press, 1981.

Russell, Bertrand. *The ABC of Relativity*. New York: New American Library, 1969.

FURTHER READING

Bodanis, David. $E=mc^2$: *A Biography of the World's Most Famous Equation*. New York: Walker, 2000.

Fairley, Peter. *Electricity and Magnetism*. Minneapolis: Twenty-First Century Books, 2007.

Fleisher, Paul. *Relativity and Quantum Mechanics: Principles of Modern Physics*. Minneapolis: Twenty-First Century Books, 2002.

McPherson, Stephanie Sammartino. *Ordinary Genius: The Story of Albert Einstein*. Minneapolis: Carolrhoda Books, 1995.

Thorne, Kip S. *Black Holes and Time Warps: Einstein's Outrageous Legacy*. New York: W. W. Norton, 1994.

WEBSITES

Albert Einstein Online
http://www.westegg.com/einstein/
This page has a wide variety of links relating to Einstein, including writings, quotes, and photos.

Einstein—Image and Impact
http://www.aip.org/history/einstein/
This online exhibit from the American Institute of Physics's Center for History of Physics is a comprehensive look at Einstein's life and work.

Einstein's Big Idea
http://www.pbs.org/wgbh/nova/einstein/
This PBS website accompanies the *Nova* program of the same name, which explores the stories behind Einstein's famous equation $E=mc^2$. The page has links to articles, interactives, and other resources.

Falling Into a Black Hole
http://casa.colorado.edu/~ajsh/schw.shtml
This Web page has animations showing the effects of relativity on space-time near a black hole.

Space Time Lab—Black Hole Interactive
http://www.universeforum.org/bh_popup_spacetimelab.htm
This interactive animation lets you visualize how space and time are distorted around a massive object like a black hole.

World Year of Physics 2005—Albert Einstein
http://www.physics2005.org/einstein.html
This page is a timeline of major events in Albert Einstein's life.

INDEX

Photo Acknowledgments

The images in this book are used with permission of: Courtesy of the Albert Einstein Archives, the Jewish National and University Library, the Hebrew University of Jerusalem, Israel, p. 5; Treasures of the National Oceanic & Atmospheric Administration Library Collection, p. 8; © Peter Willi/ SuperStock, p. 10; Library of Congress (LC-USZ62-10191), p. 13; NASA Jet Propulsion Laboratory, p. 20; © Michelson Museum, courtesy AIP Emilio Serge Visual Archives, p. 24 (left); © Case Western Reserve University, courtesy AIP Emilio Segre Visual Archives, p. 24 (right); © C. T. R. Wilson/Photo Researchers, Inc, p. 34; National Optical Astronomy Observatory/Association of Universities for Research in Astronomy/National Science Foundation, p. 41; © AIP Emilio Segre Visual Archives, Segre Collections, p. 56; © Science Museum/Science & Society Picture Library, p. 57; © Wheeler Collection, courtesy AIP Emilio Segre Visual Archives, p. 60. Diagrams and illustrations by Laura Westlund/Independent Picture Service, pp. 22, 25, 31, 33, 39, 40, 59.

Cover images by: NASA Glenn Research Center at Lewis Field (top, lower left); Illustrated London News Picture Library (lower right).